ISBN 978-1-5279-2168-9
PIBN 10902493

1 MONTH OF
FREE
READING

at

www.ForgottenBooks.com

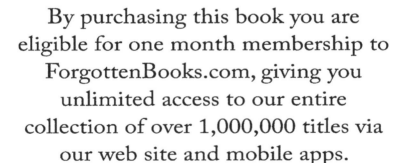

By purchasing this book you are eligible for one month membership to ForgottenBooks.com, giving you unlimited access to our entire collection of over 1,000,000 titles via our web site and mobile apps.

To claim your free month visit:

www.forgottenbooks.com/free902493

THE RELATION OF THE DENSITY TO
THE PERMEABILITY OF CONCRETE

BY

ALFRED HUGHLLYN HUNTER

THESIS

FOR THE

DEGREE OF BACHELOR OF SCIENCE

IN

CIVIL ENGINEERING

IN THE

COLLEGE OF ENGINEERING

UNIVERSITY OF ILLINOIS

PRESENTED. JUNE, 1909

UNIVERSITY OF ILLINOIS

June 1, 1909

THIS IS TO CERTIFY THAT THE THESIS PREPARED UNDER MY SUPERVISION BY

ALFRED HUGHLLYN HUNTER

ENTITLED RELATION OF THE DENSITY TO THE PERMEABILITY OF CONCRETE

IS APPROVED BY ME AS FULFILLING THIS PART OF THE REQUIREMENTS FOR THE

DEGREE OF Bachelor of Science in Civil Engineering

L. G. Parker
Instructor in Charge

APPROVED: *John P. Brooks*

HEAD OF DEPARTMENT OF Civil Engineering

144660

1.

The Relation of the Density

to the

Permeability of Concrete.

------------O------------

INTRODUCTION.

Concrete is admirably adapted to a variety of the
most important uses.   For foundations in damp and yielding soils,
subterranean masonry, retaining walls and reservoirs, under almost
every combination of circumstances. likely to be met with in
practice, it is superior to brick masonry in strength, hardness
and durability; is more economical; and in some cases is a safe
substitute for the best natural stone.

In masonry constructed of stone or brick it is almost impossible to lay the stone or brick in such a manner as to
give a monolithic structure.   Where construction of such a nature
is used for retaining walls, reservoirs, and in places where
water tends to seep through, the walls must be thicker than the
provision for equilibrium requires, in order that the amount of
water passing through may be reduced.   When concrete is used
in such construction the thickness can be reduced considerably
below that for brick or stone for an equivalent percolation,
but the wall still will be heavier than necessary to maintain
equilibrium.   Attempts have been made to relieve this fault,
but not all of them have been successful. Good results have been

obtained by care in putting a coat of neat cement on the inner
surface of the structure or by a composition of asphalt or coal
tar placed in the wall at a short distance from the inner surface.
The degree of efficiency varies largely with the character of the
labor employed so that no definite data can be obtained.   The
object of reducing the percolation is not merely to have an imper-
vious material but to increase the life of the structure. Nearly
every failure of masonry is due to the disintegration of the
exposed surface, and the presence of water on the surface tends
to make both the chemical and mechanical activities more marked.

Ordinary lime mortar absorbs between 50 and 60% of
water by volume, while mortar made from the best Portland cement
will absorb from 1/8 to 1/6 as much. Attempts have been made to
reduce this absorbing power either by means of chemical compounds
or by mechanical mixtures.

Theoretically any mixture which has no voids will
have a maximum density. At the same time a dense concrete has a
high tensile strength and a low permeability. If there is not
enough mortar to fill the voids the concrete will be weak and
porous. On the other hand an excess of mortar increases the cost,
decreases the strength and changes the permeability but in what
manner and to what extent has not yet been fully determined. With
these considerations in mind it seems feasiable to obtain a concrete of maximum density; with such a concrete to determine the
porosity; and by a process of comparison, to note the relation of

porosity to density.

To secure a mixture of maximum density the percentage
of voids must be reduced to a minimum. The proportion of voids
is independent of the size of particles, but depends upon the
gradation of the sizes, and varies with both the form and surface
of the stone. An aggregate of perpectly smooth spheres of equal
size has approximately 26% voids if closely packed. In the case
of crushed stone the voids are very much greater due to the rough
faces and projecting corners.

In order to determine the quantities of different
sizes of stone necessary to produce a maximum density, the curve
given by Fuller and Thompson in their experiment with Jerome
Park material was used. This curve has been determined by actual
experiment and in order to facilitate matters mathematical curves
were fitted to them for convenience in plotting. The curve
consists of an ellipse and a tangent straight line. The point
of tangency of the ellipse and straight line is found to be on a
vertical ordinate whose abscissa is about one tenth of the
abscissa for the maximum particles of stone.

The general equation of the ellipse about the
vertical axis is $y = \dfrac{b^2}{a^2}\sqrt{a^2 - x^2}$. This is simple for
plotting and the value of y is determined by the relation of
a and b, the major and minor axes. The values of a and b are
found to vary for different maximum sizes of stone, and for
stone having a maximum size of 0.45" , a = 0.098 inches and

4.

b = 28.39 per̯cent.

      For the data contained in this article the three curves
shown in Plate I were used. ̦ The equation of̯ *the* *first* curve was as
given above, consisting of an ellipse and a tangent line; the
second was the same as the first one except that values of a
and b were each increased 50 per̯cent. The third was simply a
straight line connecting the origin of coordinates to point
whose abscissa is 0.45 inches, and ordinate 100 per̯cent.

DESCRIPTION.

----------O----------

Cement.

Chicago A-A, Portland cement from the regular ship-
ments to the University was used in all tests. The cement was
tested according to Specifications of American Society of Civil
Engineers for all tests. The results of the above mentioned
tests are given in Table 6.

Stone.

The stone used was Joliet Limestone ranging in size
from 0.45 inches as a maximum down to dust. The specific gravity
and weight per cubic foot was determined from 10 - 12 gram
samples on an analytical balance.

The sieves used in sifting the stone were circular,
6 inches in diameter and 2 1/2 inches deep, and were identical
in form to those used in determining the fineness of cement. All
data pertaining to the commercial number and size of these seives
are given in Table 7 . The values for the diameter of the
largest size of stones passing each sieve were obtained from
tables shown by Taylor and Thompson and these checked $_\wedge$ very closely
with the ones shown in the Transactions of the American Society
of Civil Engineers for 1906.

The samples tested were molded in steel rings
6 inches in diameter 2 inches high. The rings were measured
carefully so that their volumes could be accurately determined.

Apparatus.

        The permeability apparatus used was practically the same as that used by the Structural-Materials Testing Laboratories of the United States Geological Survey at St. Louis. The apparatus is shown in *Plate* 2. It consisted of two castings 3/4 inch thick between which the disks were clamped. The disk is fitted at either end with a rubber gasket and in the later disks a thin coating of asphalt was added so as to insure water tight connections without excessive tension in the bolts connecting the castings. The water enters through a 1/2 inch nipple screwed into the upper casting. A funnel was riveted to the lower plate in order that the water passing through *could* be collected in a vessel. In the experiments conducted in the laboratory, the water was collected in a graduate or a flask fitted with a stopper through which the end of the funnel protruded. In the case of the graduate some slight error resulted from evaporation but in the closed flask no evaporation was possible.

        The materials used in the disks was proportioned according to the maximum density curves. The cement used was weighed out and then the given quantities of the various sizes of crushed stone were added. A uniform mixture was obtained by mixing thogoughly with a trowel while dry. Then the water was added and the material mixed for 5 minutes.

        The mixing being completed the concrete was placed in the iron rings in thin layers and tamped with the 2 inch round tamper shown on *Plate* 3

The disks were filled flush to the top and the surface left as nearly level as could be done with the tamper. The
specimens were then covered with a moist cloth for 24 hours, after
which they were placed in a moist chamber and left for 6 days
before being tested.

Disks numbered 1 - 11 inclusive were made according
to curve # 1 and all of them contained 11 percent of water except
disk number 2, which had 10 percent, Disks 12 and 13 were the same
as the preceding ones but contained an excess of fine material.
This fine material which was added consisted of stone dust passing
sieve #74 and was equal in amount to that passing sieve #40 and
below.   Disks No. 14 - 16 inclusive were made according to
curve #3, while disks #17 - 19 inclusive were made in accordance
with curve No. 2.

Volumetric Tests.

The density of the mixture was determined by means
of volumetric tests made upon every disk tested.

The material for the disk was mixed in a galvanized
iron pan with an ordinary five inch trowel. The mortar was placed
in the ring and thoroughly tamped.   The exact amount of the
material contained in the ring was determined by weighing the ring
and contents. From the exact measurements of the ring and the
specific gravity of the concrete, the weight of the ring when
full was calculated. In this calculation the ring is assumed

to be entirely free from voids. The difference between the actual
and calculated weights of the contents of the ring is the amount
of voids.   This number divided by the actual volume of the
disk and multiplied by 100 gives the percentage of voids.   The
results of the volumetric   tests are given, for each disk in
Table No. 4.

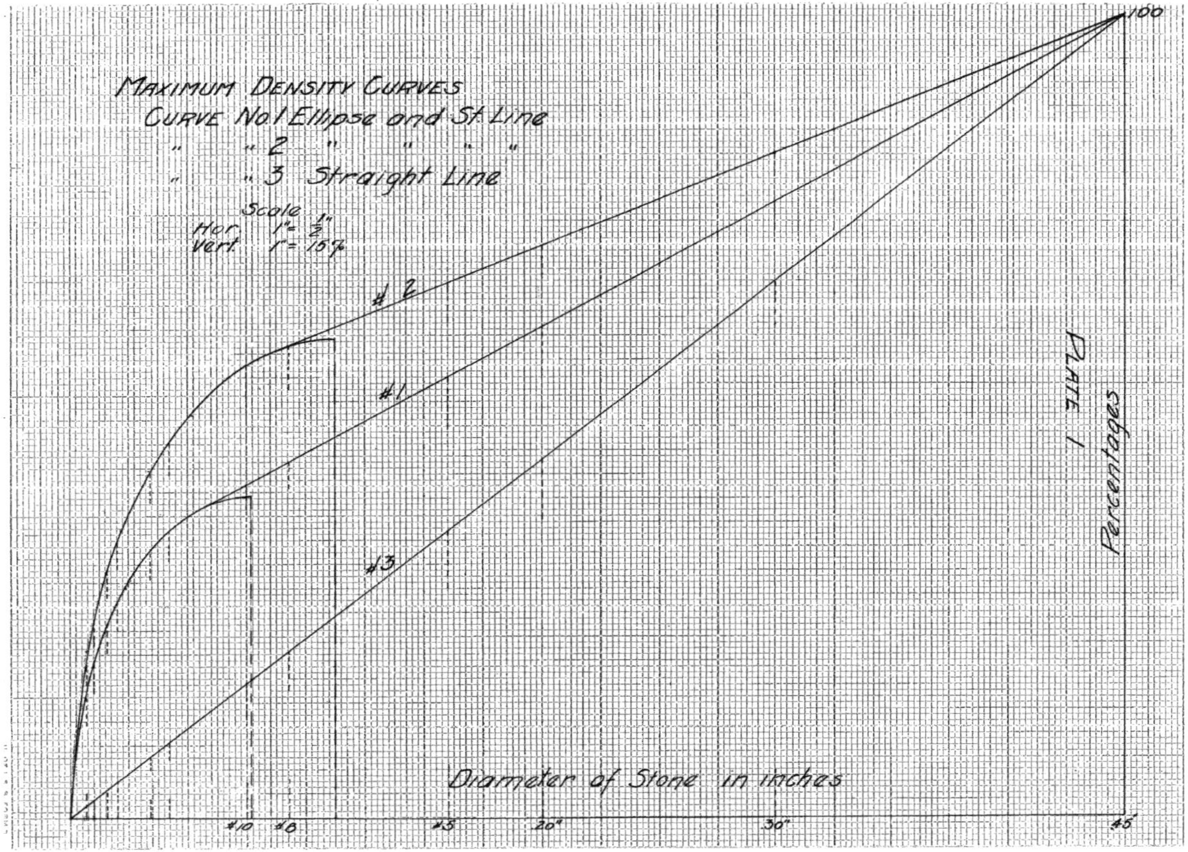

Maximum Density Curves
Curve No1 Ellipse and St Line
  "   "2   "   "   "   "
  "   "3 Straight Line

Scale
Hor: 1"= 5'
Vert: 1"= 10%

#1
#2
#3

Diameter of Stone in inches

Percentages

Plate 1

45'   30'   20'   15'   10'   5'

TABLE 1

WEIGHTS IN MIXTURE
FROM
CURVE Nº 1

Sieve No.	Percentage %	Cement	Stone	Weight Cement	Stone
Pan	7.00	7.00		175.00	
200	7.10	7.10		177.50	
100	2.00	2.00		50.00	
74	3.60	2.70	0.90	67.50	22.50
60	4.80		4.80		120.00
40	2.50		2.50		62.50
30	6.00		6.00		150.00
20	2.50		2.50		62.50
16	5.62		5.62		140.50
10	3.18		3.18		79.50
8	10.30		10.30		257.50
5	6.25		6.25		156.25
0.20"	15.65		15.65		391.25
0.30"	23.50		23.50		587.50
0.45"					
	100.00	Total Amt		470.00	2030.00

TABLE 2

WEIGHTS IN MIXTURE
FROM
CURVE Nº 3

Sieve No	Percentages %	Cement	Stone	Weights Cement	Stone	Stone
Pan	0.90	0.90				
200	0.42	0.42				
100	0.23	0.23		38.75		
74	0.78		0.78		19.50	15.10
60	1.35		1.35		33.75	26.10
40	0.77		0.77		19.25	14.95
30	3.22		3.22		80.50	62.20
20	1.71		1.71		42.75	33.10
16	7.00		7.00		175.00	135.90
10	4.37		4.37		109.25	84.45
8	14.85		14.85		371.25	286.85
5	9.04		9.04		226.00	174.60
0.20"	22.17		22.17		554.25	427.25
0.30"	33.19		33.19		829.75	640.75
0.45"						
	100.00	Add Cement 560.00				
		Total Amt 598.75				1901.25

TABLE 3

WEIGHTS IN MIXTURE
FROM
CURVE Nº2

Sieve No	Percentage			Weights	
	%	Cement	Stone	Cement	Stone
Pan	15.00	15.00		375.00	
200	2.50	2.50		62.50	
100	3.28	2.70	0.58	67.50	14.50
74	2.75		3.75		93.75
60	5.82		5.82		145.50
40	3.10		3.10		77.50
30	7.75		7.75		193.75
20	4.85		4.85		120.50
16	9.38		9.38		234.50
10	3.60		3.60		90.00
8	7.60		7.60		190.00
5	4.65		4.65		116.25
0.20"	11.40		11.40		285.00
0.30"	17.35		17.35		433.75
0.45"					
	100.00	Total Amt		505.00	1995.00

TABLE 4

PERCOLATION IN GRAMS

Disk No	Curve	Composition Wt.% Percent Stone Water	Percent Voids	Wt. grams	Age days	Grams Water Passing per Day						
						1	2	3	4	5	6	7
1	No.1	0.43	.11	10.4	2400	7	1700	133	61	38	22	19	16
2	"	"	10	13.2	2300	7	4200	100	780	218	128	104	9.5
3	"	"	11	11.0	2340	7	260	93	62	25	19	10	4
4	"	"	11	9.2	2340	7	—	198	77	28	24	14	10
5	"	"	11	12.1	2350	7	210	83	46	25	14	10	8
6	"	"	11	13.4	2340	7	258	96	51	27	18	12	10
7	"	"	11	11.6	2310	7	263	97	37	22	12	8	5
8	"	"	11	10.2	2310	7	296	65	28	15	11	8	4
9	"	"	11	11.1	2320	7	600	145	110	63	46	23	17
10	"	"	11	11.1	2320	7	415	105	38	18	10	6	2
11	No.6	"	11	9.2	2380	7	105	50	23	12	12	8	5
12	No.6	"	11	13.4	2290	7	30	95	0	0	0	0	0
13	"	"	11	10.4	2400	7	70	0	0	0	0	0	0
14	No.3	"	11	17.0	2247	7	238	169	56	49	38	40	19
15	"	"	11	13.0	2312	7	70	171	73	64	43	43	23
16	"	"	11	15.2	2253	7	441	160	54	46	33	32	18
17	No.2	"	11	11.0	2365	7	126	60	19	13	9	7	4
18	"	"	11	12.0	2317	7	96	60	9	4	1	0	0
19	"	"	11	13.5	2343	7	42	32	7	4	2	0	0

TABLE 5

PERCOLATION
IN OUNCES
PER SQUARE INCH PER DAY

Disk No.	Day 1	2	3	4	5	6	7	Total
1	6.731	0.381	0.195	0.099	0.064	0.056	0.049	7.575
2	16.860	5.250	2.508	0.565	0.375	0.305	0.291	25.974
3	1.031	0.232	0.167	0.077	0.049	0.028	0.013	1.597
4	——	0.477	0.208	0.086	0.062	0.039	0.032	0.904
5	0.708	0.206	0.119	0.070	0.039	0.028	0.028	1.198
6	0.869	0.239	0.132	0.076	0.051	0.034	0.035	1.436
7	0.788	0.236	0.096	0.067	0.031	0.023	0.023	1.264
8	0.877	0.151	0.076	0.046	0.029	0.023	0.018	1.230
9	1.928	0.337	0.284	0.163	0.135	0.062	0.048	2.957
10	1.334	0.272	0.103	0.053	0.029	0.017	0.006	1.814
11	0.337	0.129	0.062	0.031	0.035	0.022	0.014	0.630
12	0.084	0.291	0	0	0	0	0	0.375
13	0.197	0	0	0	0	0	0	0.197
14	0.975	0.368	0.164	0.132	0.117	0.003	0.058	1.887
15	0.277	0.348	0.214	0.172	0.132	0.103	0.077	1.323
16	1.750	0.372	0.158	0.124	0.101	0.083	0.055	2.643
17	0.502	0.131	0.056	0.035	0.028	0.018	0.012	0.782
18	0.381	0.130	0.026	0.011	0.003	0	0	0.551
19	0.167	0.069	0.021	0.011	0.006	0	0	0.274

Table 6
Tests
of
Chicago A-A Portland
Cement

FINENESS				
Sieve	Retained	Passing	% Retained	% Passing
50				
74	27	973	2.7	97.3
100	42	931	4.2	93.1
200	202	729	20.2	72.9
Pan	729		72.9	

Tensile Strength

7 Days		28 Days	
Neat	1-3	Neat	1-3
610	91	815	168
662	119	775	210
750	84	815	172
690	50	815	156
714	89	785	172
740	—	865	166
Av 694	87	812	174

Specific Gravity

No.	Wt.	Vol.	Sp. Gr.
1	64	20.41	3.13
2	64	20.14	3.17
	Average		3.15

9

CONCLUSION.

-------O-------

      As has been stated, previous tests indicate that the greatest strength from any given percentage of cement was obtained when the concrete was of the greatest possible density, that is having the least percentage of voids, and further that the greatest density was obtained when all the materials were proportioned so as to give a regular mechanical analysis curve approaching a parabola. It also appeared probable that concrete of the greatest density would be the the least permeable by water.

      With the foregoing conclusions in mind it was thought profitable to test specimens made by a maximum density curve and compare the results so obtained to the samples made from mixture having a known variation from that of maximum density.

      Test specimens from 1 - 10 inclusive were made from mixtures obtained from curve No. 1 and were found to have an average of 11.1 per cent of voids and 1.259 ounces per square inch per week of percolation. Disk No. 2 was not included in this average as 10 instead of 11 per cent of water had been used. and the percolation which resulted was very high. This lower percentage of water in disk No. 2 seemed to cause the cement to adhere to the stone particles in a manner similar to that shown in ordinary dough and no amount of tamping caused the water to flush to the surface. The other examples have considerable

variation in the amount of water passing but no so much but that
any one could be classified as fairly representative of the mixture.   Disk No's 11 - 13 inclusive were the same as the ones
just mentioned but containing an excess of fine material, the
amount of which is given in a previous paragraph. The average
per cent of voids and percolation in ounces per square inch per
week was found to be 11.0 and 0.286 respectively. The percentage
of voids remained very nearly the same but the amount of percolation was reduced to a great extent, so much so that in two of
the samples Nos. 12 and 13 the evaporation equalled the percolation for the last 5 days. This result although not conclusive
seems to bear out the statement of other experimentors , that
an excess of fine material reduces the amount of water passing
to a lower value than that obtained from the maximum density
curve samples.

            Disk Nos. 17 - 19 inclusive are made from curve No.2
and gives results as follows; voids 12.26 per cent ; percolation
0.536 ounces per square inch per week. This small amount of
percolation can be explained by the fact that there was consid-
erable more fine material in the mixture than that obtained from
the maximum density curve, and at the same time the decrease
in the amount of the larger size of stone caused a decrease in
the density.

            Disk Nos 14 to 16 inclusive were made according to
data obtained from the straigh line curve shown at 3 Plate 1,

and had 15.7 per cent voids and a percolation of 1.952 ounces
per square inch per week. These values when compared to those
of curves No's. 1 and 2 are found to be much in excess both with
regard to percolation and to the percentages of voids.

From the results obtained above the permeability
is found to vary inversely as the density of the mixture and
that an excess of fine material changes the density but slightly
and reduces the permeability. Also that in tests of this
character better results may be obtained by an excess of water in
making the disk. The results while not conclusive, on account
of the small number of tests, give data of such a character
that would indicate that mixtures which approach the maximum
density curve are those which contain an excess of fine material
will be the most satisfactory where a low permeability is requir-
ed.

In further experiments of a similar nature effort
should be made to secure uniform water pressure, for in the
experiment described above the pressure was found to range
from 45 lb. per square inch as normal to occassional pressures
which would blow out the asphalt between the gasket and the
disk.   The effect of this variation was noticeable in the
daily readings of some of the specimens.

CPSIA information can be obtained
at www.ICGtesting.com
Printed in the USA
BVHW050050061118
532207BV00023B/3208/P

9 781527 921689